INDUSTRY STANDARD OF THE PEOPLE'S REPUBLIC OF CHINA

Code for Design of Environmental Protection for Railway

TB 10501-2016

Prepared by: China Railway SIYUAN Survey and Design Group Co., Ltd.
Approved by: National Railway Administration
Effective date: October 1, 2016

China Railway Publishing House

Beijing 2019

图书在版编目(CIP)数据

铁路工程环境保护设计规范:TB 10501-2016:英文/中华人民共和国国家铁路局组织编译.—北京:中国铁道出版社,2019.1
ISBN 978-7-113-24923-6

Ⅰ.①铁… Ⅱ.①中… Ⅲ.①铁路工程-环境保护-设计规范-中国-英文 Ⅳ.①X731-65

中国版本图书馆 CIP 数据核字(2018)第 202228 号

Chinese version first published in the People's Republic of China in 2016
English version first published in the People's Republic of China in 2019
by China Railway Publishing House
No. 8, You'anmen West Street, Xicheng District
Beijing, 100054
www. tdpress. com

Printed in China by Beijing Hucais Culture Communication Co., Ltd.

© 2016 by National Railway Administration of the People's Republic of China

All rights reserved. No part of this publication may be reproduced or transmitted in any form or by any means, electronic or mechanical, including photocopying, recording, or by any information storage and retrieval systems, without the prior written consent of the publisher.

This book is sold subject to the condition that it shall not, by way of trade or otherwise, be lent, resold, hired out or otherwise circulated without the publisher's prior consent in any form of binding or cover other than that in which it is published and without a similar condition including this condition being imposed on the subsequent purchaser.

ISBN 978-7-113-24923-6

Introduction to the English Version

The translation of this Code was made according to Railway Engineering and Construction Development Plan of the Year 2017 (Document GTKFH [2017] No. 185) issued by National Railway Administration for the purpose of promoting railway technological exchange and cooperation between China and the rest of the world.

This Code is the official English language version of TB 10501-2016. In case of discrepancies between the original Chinese version and the English translation, the Chinese version shall prevail.

Planning and Standard Research Institute of National Railway Administration is in charge of the management of the English translation of railway industry standard, and China Railway Economic and Planning Research Institute Co., Ltd. undertakes the translation work. Sichuan Lanbridge Information Technology Co., Ltd. and China Railway Engineering Consulting Group Co., Ltd. provided great support during translation and review of this English version.

Your comments are invited and should be addressed to China Railway Economic and Planning Research Institute Co.,Ltd.,29B, Beifengwo Road, Haidian District, Beijing, 100038 and Planning and Standard Research Institute of National Railway Administration, Building B, No. 1 Guanglian Road, Xicheng District, Beijing, 100055.

Email: jishubiaozhunsuo@126.com

The translation was performed by Ma Liya, Dong Suge, Chai Guanhua, Chen Guocai.

The translation was reviewed by Chen Shibai, Wang Lei, Yang Quanliang, Liu Dalei, Han Lihe, Dong Mai, Wang Yajun.

Notice of National Railway Administration on Issuing the English Version of Four Railway Standards including *Code for Design of Environmental Protection for Railway*

Document GTKF [2019] No. 6

The English version of *Code for Design of Environmental Protection for Railway* (TB 10501-2016), *Code for Design of Energy Conservation for Railway* (TB 10016-2016), *Code for Design of Fire Prevention for Railway* (TB 10063-2016) and *Code for Design of Concrete Structures of Railway Bridge and Culvert* (TB 10092-2017) is hereby issued. In case of discrepancies between the original Chinese version and the English version, the Chinese version shall prevail.

China Railway Publishing House is authorized to publish the English version.

National Railway Administration
January 28, 2019

Notice of National Railway Administration on Issuing Railway Industry Standard (Engineering and Construction Standard Batch No. 4, 2016)

Document GTKF [2016] No. 27

Code for Design of Environmental Protection for Railway (TB 10501-2016) is hereby issued and will come into effect on October 1, 2016. *Code for Design of Environmental Protection for Railway* (TB 10501-98) is withdrawn.

China Railway Publishing House is authorized to publish this Code.

National Railway Administration
June 29, 2016

Foreword

Based on *Code for Design of Environmental Protection for Railway* (TB 10501-98), this Code is prepared by summarizing and drawing on the experience gained over recent years in design of environmental protection for railway, referring to the environmental protection design standards at home and abroad and extensively soliciting opinions from all sides, as well as through review and revision.

The Code consists of 10 chapters, namely General Provisions, Terms, Route Selection and Site Selection Based on Environmental Protection Design, Ecological Environment Protection, Prevention and Control of Noise Pollution, Prevention and Control of Vibration Pollution, Prevention and Control of Water Pollution, Prevention and Control of Air Pollution, Prevention and Control of Solid Waste Pollution, Prevention and Control of Electromagnetic Pollution.

The main revisions are as follows:

1. Addition of the provisions on waste recycling and reclamation as well as the layout of pollution control facilities, and specification of the requirements on environmental protection and soil and water conservation during the construction period in "General Provisions".

2. Addition of a new chapter titled "Terms".

3. Addition of a new chapter titled "Route Selection and Site Selection Based on Environmental Protection Design", as well as specification of basic principles for route selection and site selection of railway engineering.

4. Addition of the requirements for "Design for Biological Conservation and Habitat Conservation", "Design for Land Resources Conservation" and "Green and Landscape Design" in Chapter "Ecological Environment Protection".

5. Division of Chapter "Prevention and Control of Noise and Vibration Pollution" in the previous code into two chapters, namely "Prevention and Control of Noise Pollution" and "Prevention and Control of Vibration Pollution" in this Code.

6. Specification of the requirements on sound barrier arrangement and requirements on acoustic and structural design in Chapter "Prevention and Control of Noise Pollution".

7. Specification of requirements on prevention and control of air pollution during the construction period and the operation period in Chapter "Prevention and Control of Air Pollution".

We would be grateful if anyone finding the inaccuracy or ambiguity while using this Code would inform us and address the comments to China Railway SIYUAN Survey and Design Group Co., Ltd. (No. 745, Heping Avenue, Yangyuan Street, Wuchang District, Wuhan, Hubei Province, 430063) and China Railway Economic and Planning Research Institute Co., Ltd. (No. 29B Beifengwo Road, Haidian District, Beijing, 100038) for the reference of future revisions.

The Technology and Legislation Department of National Railway Administration is responsible for the interpretation of this Code.

Prepared by:
China Railway SIYUAN Survey and Design Group Co., Ltd.

Drafted by:
Wang Zhonghe, Tian Chao, Gong Ping, Zhang Weihong, Xu Ping, Gao Zhiliang, Shi Juan, Lu Shaofei, Wu Fang, Wang Zhengang and Zhang Liangtao.

Reviewed by:
Chen Jiandong, Li Jia, Zhang Songyan, Liao Jianzhou, Liu Yan, Sang Cuijiang, Ru Xu, Wang Zhehao, Han Yan, Xin Siyuan, Song Jun, Tian Yang, Liu Xun, Chen Jun, Yang Sibo, Li Yaozeng, Zhao Liuhui, Xia Xianfang, Zhu Zhengqing, Su Weiqing, Chen Shulian, Guo Liancheng, Chen Jie, Li Cangsong, Chen Gang, Hu Bo and Yuan Wenzhong.

Contents

1 General Provisions ……………………………………………………………………… 1
2 Terms ……………………………………………………………………………………… 2
3 Route Selection and Site Selection Based on Environmental Protection Design ………… 3
4 Ecological Environmental Protection …………………………………………………… 4
    4.1 Design for Biological Conservation and Habitat Conservation ……………………… 4
    4.2 Design for Land Resources Conservation ………………………………………… 4
    4.3 Design for Water and Soil Conservation …………………………………………… 4
    4.4 Green and Landscape Design ……………………………………………………… 5
5 Prevention and Control of Noise Pollution ……………………………………………… 7
    5.1 General Requirements ………………………………………………………………… 7
    5.2 Design for Prevention and Control of Noise Pollution During Construction Period …… 7
    5.3 Design for Prevention and Control of Noise Pollution During Operation Period ……… 7
6 Prevention and Control of Vibration Pollution ………………………………………… 9
    6.1 General Requirements ………………………………………………………………… 9
    6.2 Design for Prevention and Control of Vibration Pollution During Construction
        Period ………………………………………………………………………………… 9
    6.3 Design for Prevention and Control of Vibration Pollution During Operation Period …… 9
7 Prevention and Control of Water Pollution ……………………………………………… 10
    7.1 General Requirements ………………………………………………………………… 10
    7.2 Design for Prevention and Control of Wastewater Pollution During Construction
        Period ………………………………………………………………………………… 10
    7.3 Design for Prevention and Control of Wastewater Pollution During Operation
        Period ………………………………………………………………………………… 10
8 Prevention and Control of Air Pollution ………………………………………………… 11
    8.1 General Requirements ………………………………………………………………… 11
    8.2 Design for Prevention and Control of Air Pollution During Construction Period ……… 11
    8.3 Design for Prevention and Control of Air Pollution During Operation Period ………… 11
9 Prevention and Control of Solid Waste Pollution ……………………………………… 12
    9.1 General Requirements ………………………………………………………………… 12
    9.2 Design for Prevention and Control of Solid Waste Pollution During Construction
        Period ………………………………………………………………………………… 12
    9.3 Design for Prevention and Control of Solid Waste Pollution During Operation
        Period ………………………………………………………………………………… 12
10 Prevention and Control of Electromagnetic Pollution ………………………………… 13
Words Used for Different Degrees of Strictness ………………………………………… 14

1 General Provisions

1.0.1 The Code is prepared with the view to implementing the national laws, regulations and policies on environmental protection and to unifying the design standards for environmental protection of railway engineering.

1.0.2 This Code is applicable to the design of environmental protection for new railway or upgrading of existing railway.

1.0.3 The design of environmental protection shall follow the principle of "protection first, prevention first and comprehensive treatment", so as to avoid or reduce the adverse environmental impact caused by railway projects. Environmental protection measures shall be in accordance with local conditions, technically feasible and economically rational.

1.0.4 Technologies, processes, equipment or materials prohibited by the government must not be used in the design of railway engineering.

1.0.5 Design of railway engineering shall comply with relevant national regulations on waste recycling and reclamation.

1.0.6 Pollutants resulted from railway engineering shall meet the requirements of national or local discharge standards and shall meet the requirement of control over the total amount of major pollutants.

1.0.7 Pollution control of railway station areas shall be designed in conjunction with the layout and planning of local pollution control facilities.

1.0.8 The facilities for pollution prevention and control as well as for ecological protection during railway engineering and main works shall be designed simultaneously in accordance with requirements specified in approved environmental impact assessment documents.

1.0.9 The environmental protection measures as well as soil and water conservation requirements during the construction period shall be proposed according to the environmental characteristics of the project area in construction organization design.

1.0.10 In addition to this Code, the environmental protection design of railway engineering shall also comply with relevant provisions specified in current national standards.

2 Terms

2.0.1 Ecologically sensitive region

The ecologically sensitive region includes special ecologically sensitive region and major ecologically sensitive region.

The special ecologically sensitive region refers to the area that has extremely important ecological service function, extremely vulnerable ecosystem or serious ecological problems, and where occupation, damage or destruction causes serious and unpreventable ecological impacts and ecological function is impossible to be restored or replaced, including natural reserves and world cultural and natural heritage sites.

The major ecologically sensitive region refers to the area that has relatively important ecological service function or relatively vulnerable ecosystem, and where occupation, damage or destruction causes relatively serious but preventable ecological impacts and ecological function can be restored or replaced, including scenic spots, forest parks, geoparks, important wetlands, primitive natural forests, naturally concentrated distribution areas for rare and endangered wildlife, natural spawning grounds of important aquatic life, feeding ground, wintering ground, breeding migration passage and natural fishery.

2.0.2 Wildlife passage

Buildings or structures built across railway line for safe passage of wildlife to protect them and enable their migration.

2.0.3 Land reclamation

Rehabilitation measures taken for temporary land damaged by construction activities to make it arable again.

2.0.4 Vegetation measures for water and soil conservation

Measures such as planting and seeding, based on the biological principles, taken to prevent and control water and soil erosion caused by construction activities and also to protect, improve and rationally utilize land and water resources.

2.0.5 Engineering measures for water and soil conservation

Engineering facilities built based on the engineering principles, to prevent and control water and soil erosion caused by construction activities and also to protect, improve and rationally utilize land and water resources.

2.0.6 Habitat

Environmental conditions of areas for inhabitation, breeding, foraging and migration of individuals, population or community, including necessary living conditions and other ecological factors beneficial to the creatures.

3 Route Selection and Site Selection Based on Environmental Protection Design

3.0.1 In the process of route selection and site selection for railway engineering, the core areas and buffer zones of the natural reserves, the core areas of the scenic spots, the world cultural and natural heritage sites and the Class I protection areas of drinking water source must be bypassed. Production facilities discharging pollutants shall not be arranged in the Class II protection areas of drinking water source. Production facilities polluting environment or damaging resource or landscape shall not be arranged in the experimental areas of natural reserves. The soil (stone) borrow pit shall not be located in the area with collapse and landslide risk, area prone to mudslides as well as watercourse and lake management area determined by the local people's government at the county level and above.

3.0.2 In the process of route selection and site selection for railway engineering, the following places should be bypassed, including the experimental areas of natural reserves, other tourist attractions beyond the core area of the scenic spots, forest parks, geoparks, important wetlands, natural forests, naturally concentrated distribution areas for rare and endangered wildlife, natural spawning grounds of important aquatic life, feeding ground, wintering ground, breeding migration passage, natural fishery and other protection areas beyond the Class I protection areas of drinking water source.

3.0.3 If the environmental sensitive regions specified in Article 3.0.2 cannot be bypassed, the construction plan shall be determined through comprehensive comparison and selection, and corresponding measures shall be taken for environmental protection and ecological restoration.

3.0.4 The route selection and site selection for railway engineering shall be coordinated with the planning of cities and towns and program of environmental protection.

3.0.5 In the process of route selection and site selection for railway engineering, the following places shall be bypassed, including the major areas for prevention and control of water and soil erosion, areas prone to mudslides, areas with collapse and landslide risks, areas prone to severe water and soil erosion and ecological deterioration as well as soil and water conservation monitoring stations and key experimental areas in the national soil and water conservation monitoring network.

4 Ecological Environmental Protection

4.1 Design for Biological Conservation and Habitat Conservation

4.1.1 In case of railway passing through woodland, grassland and desertification area, the following protection measures shall be taken:

1 When railway passing through woodland and grassland, the land occupation and deforestation area of the woodland and grassland shall be controlled strictly, and schemes and measures for vegetation protection and restoration during the construction period as well as for fire protection and isolation during the operation period shall be provided.

2 When railway passing through desertification area, the eco-environmental protection measures shall focus on the engineering measures and shall be supplemented by vegetation measures.

4.1.2 When the railway line affects the migration of wild animals under protection, passages shall be built to facilitate the migration activities of wild animals.

4.1.3 When the railway construction affects the ancient and rare trees, measures such as providing enclosures and fences, transplantation, protection or bypassing shall be taken.

4.1.4 If tunnel project construction is likely to result in groundwater leakage or seriously affects the surface ecological environment and water supply for production and living of residents, the measures for water resource protection and water pollution prevention and control shall be taken according to the results of geological prediction.

4.2 Design for Land Resources Conservation

4.2.1 When the railway line goes through the prime farmland, the scheme of bridge is preferred.

4.2.2 In the case of farmland, garden, woodland and grassland occupied by railway line, the schemes and measures for stripping, preservation and utilization of the surface cultivated soil shall be provided.

4.2.3 The method of cut-fill balance shall be used in railway project so as to reduce the scale and number of borrow pits and spoil grounds.

4.2.4 The temporary land utilization during the construction period of the railway project shall be in accordance with the principle of "combination of permanent land utilization with temporary land utilization", and the barren land and degraded land should be used to avoid or reduce farmland occupation.

4.2.5 The reclamation of the land temporarily used by the railway project shall be designed in accordance with the determined land reclamation scheme, and the reclamation shall comply with the provisions specified in *Completion Standards on Land Reclamation Quality* (TD/T 1036).

4.3 Design for Water and Soil Conservation

4.3.1 Vegetation measures or vegetation measures in combination with engineering measures shall be used in the railway project, so as to prevent and control water and soil erosion; the plant-species selection principles and the planting technology requirements shall be clearly specified.

4.3.2 Regarding the works including earthworks, bridge, tunnel and station/yard of the railway project, their drainage systems shall be designed in such a way that the surface runoff can be smoothly discharged into the local drainage system according to the landform.

4.3.3 Slope protection measures of earthworks shall be selected and used according to the features of soil (stone) of the slope, gradient and height of slope, climate, etc.; if the safety and stability of the works can be ensured, the engineering measures in combination with vegetation measures shall be adopted.

4.3.4 Mud and drilling slag generated during bridge construction shall be transported to the designated site for disposal after being treated in the sedimentation tank.

4.3.5 The front slope and side slope of the tunnel portal should be protected by engineering measures in combination with vegetation measures.

4.3.6 The prevention and control of water and soil erosion for temporary railway works shall comply with the following provisions:

1 Drainage ditches shall be provided in the vicinity of construction sites of large temporary works, and a sedimentation tank shall be set at the outlet in the downstream area.

2 Exposed ground surfaces and temporarily stacked slag/soil shall be provided with temporary coverage measures.

3 The site shall be provided with measures such as land treatment, slope protection and landscaping for restoration and utilization. In case of the site occupying farmland, land reclamation shall be carried out.

4 Access roads shall be provided with protection measures according to the soil and water conservation requirements.

4.3.7 The prevention and control of water and soil erosion for soil (stone) borrow pit and spoil ground shall comply with the following provisions:

1 Spoils shall be utilized comprehensively. Spoils to be discarded shall be stored in special storage yards and provided with protection measures.

2 On the spoil ground, the spoil retaining structures shall be built in advance according to the conditions of landform, geology and hydrology.

3 The level and slope site shall be designed with good drainage systems which shall be smoothly connected to the local drainage ditches or channels.

4 The level and slope ground shall be protected by vegetation measures or engineering measures, or by vegetation measures in combination with engineering measures.

4.3.8 In the case of railway located in dust bowl, the construction operation range shall be determined according to the physical characteristics of wind and sand, vegetation coverage and other natural conditions; appropriate measures for soil and water conservation such as engineering measures, vegetation measures, or vegetation measures in combination with chemical measures shall be used.

4.4 Green and Landscape Design

4.4.1 In the case of railway passing through the area with high landscape requirement, vegetation measures shall be taken so as to coordinate with the surrounding landscape.

4.4.2 Green design of the areas available for green landscape in railway projects shall be coordinated with the surrounding environment and shall comply with the following provisions:

1 The method of "combining shrubs and herbs with priority given to shrubs" should be used for green landscape of railway earthworks slope.

2 Green landscape under the railway bridge shall focus on grass planting, and the method of planting shrubs or the method of combining shrubs with herbs should be adopted on both sides.

3 The green landscape for side slope and front slope of the tunnel portal shall focus on grass planting, and the method of planting shrubs or the method of combining shrubs with herbs should be adopted on the top of the open cut tunnel.

4 Ornamental evergreen shrubs, arbors and flowers should be used for the greening of office areas of station/yard. The method of mixed planting of evergreen trees, deciduous trees, broad-leaved trees and needle-leaved trees should be used for the greening of production areas. Fences and enclosures of the station area should be covered with lianas or high hedgerows.

5 If there are conditions for greening the level and slope site of the soil (stone) borrow pit and spoil ground, the method of grass planting and the method of combining shrubs with herbs shall be used.

5 Prevention and Control of Noise Pollution

5.1 General Requirements

5.1.1 In the design for prevention and control of railway noise pollution, the noise-sensitive building or the concentrated area of noise-sensitive buildings shall be regarded as the noise-sensitive point.

5.1.2 In the design for prevention and control of railway noise pollution, the engineering measures or comprehensive prevention and control measures shall be proposed from the aspects such as reducing the noise source intensity, blocking the propagation path and protecting the noise-sensitive point.

5.2 Design for Prevention and Control of Noise Pollution During Construction Period

5.2.1 In construction organization design, the measures for reducing noise effect of construction machinery near the noise-sensitive buildings shall be proposed. The construction noise shall comply with the provisions specified in *Emission Standard of Environment Noise for Boundary of Construction Site* (GB 12523).

5.2.2 In construction organization design, measures for noise control of blasting operation of railway project near the noise sensitive area shall be proposed in accordance with provisions specified in *Safety Regulations for Blasting* (GB 6722).

5.3 Design for Prevention and Control of Noise Pollution During Operation Period

5.3.1 Railway noise emission shall comply with provisions specified in *Emission Standards and Measurement Methods of Railway Noise on the Boundary Alongside Railway Line* (GB 12525) and *Emission Standard for Industrial Enterprises Noise at Boundary* (GB 12348).

5.3.2 The design for prevention and control of noise pollution shall comply with the following provisions:

    1 General equipment shall meet the requirements of low-noise equipment specified in relevant national or industrial standards.

    2 The loudspeakers at passenger station and marshalling station, water resistance test platform at diesel locomotive depot, hump retarder and other high-noise equipment at station/depot (post) shall be provided with noise reduction measures according to the characteristics of noise source and the distribution of surrounding noise-sensitive buildings.

    3 The railway track should be provided with continuous welded rail and heavy rail, and may be provided with noise reduction measures.

5.3.3 In order to block noise propagation path, the measures such as sound barriers or dense-planting tree belts should be used in the design for prevention and control of noise pollution.

5.3.4 In order to protect noise-sensitive point, the design for prevention and control of noise pollution shall comply with the following provisions:

    1 Noise-sensitive building should be provided with measures such as strengthening sound insulation or installation of soundproof doors and windows.

    2 Regarding the noise-sensitive building subject to the railway noise pollution, if the technical and economic evaluation demonstrates that its service function requirement cannot be met, measures such as building function replacement or relocation shall be taken.

5.3.5 Conditions for setting sound barriers shall comply with the following provisions:

1 In case there are at least 10 households within the area where the longitudinal continuous length along the railway line is 100 m and the dimension from the centerline of the outer track of the railway line is 80 m, with the railway noise emission exceeding the limit value specified in the current standard - *Emission Standards and Measurement Methods of Railway Noise on the Boundary Alongside Railway Line* (GB 12525), the measures of sound barriers shall be taken.

2 In case there are schools and hospitals (nursing house and gerocomium) within the area where the dimension from the centerline of the outer track of the railway line is 80 m, with the railway noise emission exceeding the limit value specified in the current standard-*Emission Standards and Measurement Methods of Railway Noise on the Boundary Alongside Railway Line* (GB 12525), the measures of sound barriers shall be taken.

5.3.6 The length and height of the sound barrier shall be calculated and determined based on the calculation models of infinite-length linear sound source and finite-length sound barrier and in accordance with the requirements on noise reduction.

5.3.7 The length of sound barrier shall be the distribution length of the noise-sensitive building along the railway line plus the additional lengths at both ends, and the additional lengths shall be determined by acoustic calculation.

5.3.8 The design for sound barriers shall comply with the following provisions:

1 The design of sound barrier structure shall comply with provisions of the national and railway industrial standards, and shall be beneficial to the repair and maintenance of sound barriers.

2 The sound barriers should be built near the sound source. The sound barriers for embankment should be built on the shoulders in accordance with the operation requirements of track maintenance. The sound barriers for bridge shall be built on the place where there are handrails of the operation passage.

3 The sightseeing effect on passengers within the running train shall be taken into account during design of sound barriers, and the outside landscape shall be coordinated with the surrounding environment, and shall not produce glare or reflected light.

4 The sound barriers shall be provided with measures against sound leakage.

5 The technical performance of the acoustic elements of the sound barrier shall comply with provisions specified in *Technique Requirements and Measurement of Acoustic Elements of Railways Sound Barrier* (TB/T 3122).

5.3.9 If the tree belt is used for noise reduction, the shade-tolerant and evergreen indigenous plants suitable for dense planting shall be selected according to the natural conditions, and arbors, shrubs and herbs shall be planted in a mixed manner.

5.3.10 The design of soundproof doors and windows of noise-sensitive building shall comply with the requirements of airborne-sound insulation of external window facing the trunk road, specified in *Code for Design of Sound Insulation of Civil Buildings* (GB 50118). The technical performance of the soundproof window used in design shall comply with the provisions specified in *Windows for Sound Insulation* (HJ/T 17).

5.3.11 In viaduct-type station, the architectural and structural measures for noise reduction shall be taken for the station building which is located below or above the track level.

6 Prevention and Control of Vibration Pollution

6.1 General Requirements

6.1.1 In the design for prevention and control of vibration pollution in railway projects, the vibration-sensitive building or the area with special requirements on the vibration environment quality shall be regarded as the vibration-sensitive point.

6.1.2 In the design for prevention and control of vibration pollution in railway projects, the engineering measures or comprehensive prevention and control measures shall be proposed from the aspects such as reducing the vibration source intensity, blocking the propagation path and vibration isolation for buildings.

6.2 Design for Prevention and Control of Vibration Pollution During Construction Period

6.2.1 In construction organization design, the measures for reducing vibration effect of construction machinery near the vibration-sensitive buildings shall be proposed. The construction vibration shall comply with provisions specified in *Standard of Environmental Vibration in Urban Area* (GB 10070) and relevant standards.

6.2.2 In construction organization design, measures for vibration control for blasting operation near the vibration-sensitive buildings shall be proposed in accordance with provisions specified in *Safety Regulations for Blasting* (GB 6722).

6.3 Design for Prevention and Control of Vibration Pollution During Operation Period

6.3.1 The design for prevention and control of vibration pollution for railway shall comply with provisions specified in *Standard of Environmental Vibration in Urban Area* (GB 10070) and relevant national or local standards.

6.3.2 The prevention and control of vibration pollution for railway shall comply with the following provisions:

  1 The low vibration equipment in line with relevant national or industrial standards shall be selected.

  2 Continuous welded rail and heavy rail should be used.

  3 The distance between the vibration source and the vibration-sensitive point should be increased.

  4 Vibration mitigation measures shall be applied to track and bridge structures as well as equipment foundation.

  5 For the vibration-sensitive building subject to the railway vibration pollution, if the technical and economic evaluation demonstrates that its service function requirement cannot be met, measures such as function replacement or relocation shall be taken.

7 Prevention and Control of Water Pollution

7.1 General Requirements

7.1.1 Production and domestic sewage of railway should be treated in a centralized manner, and shall be discharged in an orderly manner. The discharge outlets shall be set in accordance with provisions specified in relevant national or local standards.

7.1.2 Initial rainwater, leaching water from the ground contaminated by toxic and harmful substances and the area for storage and stacking of toxic and harmful substances, and wastewater containing all kinds of toxic and harmful substances shall be collected and treated with a collection system with leak-proof measures.

7.1.3 The drainage of production and domestic sewage of railway and sewage produced during the construction period shall comply with provisions specified in *Integrated Wastewater Discharge Standard* (GB 8978) or local sewage discharge standards.

7.1.4 It is strictly forbidden to discharge wastewater into seepage wells, seepage pits, fissures or karst caves.

7.2 Design for Prevention and Control of Wastewater Pollution During Construction Period

7.2.1 Wastewater resulted from the construction camp, construction site, girder fabrication (storage) yard, track slab fabrication yard, concrete batching plant and other large temporary works of the railway shall be treated.

7.2.2 In case of railway bridges crossing sensitive water bodies, the measures for collection and onshore treatment of mud, waste oil and other contaminants produced in construction shall be taken.

7.2.3 Wastewater produced in tunnel construction shall be treated.

7.3 Design for Prevention and Control of Wastewater Pollution During Operation Period

7.3.1 The production sewage of railway shall be treated, and the treated sewage should be recycled.

For the oily sewage resulted from repair and washing of locomotives and rolling stocks, floating oil should be pre-treated at the workshop discharge outlet. Acid wastewater from the battery room should be pre-treated via acid-base neutralization at the workshop discharge outlet.

7.3.2 The fecal sewage from the passenger train shall be treated.

7.3.3 Ground prone to oil pollution such as oil tank area and operation area of oil unloading siding shall be hardened integrally, collecting ditches shall be built around those areas and ground washing water and initial rainwater shall be collected and treated.

7.3.4 Dust removal wastewater resulted from the railway quarry, dedicated coal yard and coal unloading siding shall be treated and recycled.

7.3.5 The initial rainwater in the bulk goods yard, dedicated coal yard and coal unloading siding shall be treated and should be recycled.

7.3.6 Wastewater containing Type I pollutants such as Lead and Cadmium must be collected and treated in the workshop.

8 Prevention and Control of Air Pollution

8.1 General Requirements

8.1.1 Centralized heating should be adopted for railway station/depot (post) and living zones. For buildings along the railway line without conditions for centralized heating, the energy-saving heating mode should be adopted.

8.1.2 The boiler room should be located at the place where the pollution coefficient is minimum for the leeward environment sensitive point.

8.1.3 The air pollutants discharged from the boilers and the smoke, dust or harmful gases generated from the process rooms shall comply with the provisions specified in *Emission Standard of Air Pollutants for Boiler* (GB 13271) and *Integrated Emission Standard of Air Pollutants* (GB 16297) or local air pollutant emission standards.

8.1.4 The comprehensive freight yards for railway such as bulk goods areas or bulk goods storage yard should be located on the windward side of the overall minimum-frequency wind direction of the urban area.

8.1.5 Platforms, stables or sheds, watering places and other auxiliary facilities for livestock loading and unloading services shall be away from the receiving-departure track for passenger trains.

8.2 Design for Prevention and Control of Air Pollution During Construction Period

8.2.1 Measures for prevention and control of dust pollution shall be taken for the railway construction sites in the urban planning area. Prevention and control of dust pollution resulted from construction shall comply with provisions specified in *Technical Specifications for Urban Fugitive Dust Pollution Prevention and Control* (HJ/T 393).

8.2.2 The exhaust emission of construction machinery with compression ignition engine shall comply with provisions specified in relevant national or local exhaust emission standards for construction machinery.

8.3 Design for Prevention and Control of Air Pollution During Operation Period

8.3.1 The coal storage yard and slag storage yard of the boiler room shall be provided with dust suppression measures.

8.3.2 The catering service facilities for railway should be provided with clean energy and shall be equipped with high-efficiency oil fume purification facilities.

8.3.3 Processes and equipment resulting in smoke, dust and harmful gases shall be provided with dust removal and purification devices.

8.3.4 The dust prevention and control measures such as wind prevention and dust suppression shall be taken for the comprehensive freight yards for railway according to the category of goods, storage scale and environmental conditions. The dust suppression measures shall be taken for the loading and unloading operation for bulk goods.

8.3.5 The dust suppression measures shall be taken for freight trains for bulk coal and ores prone to producing dust with wind.

9 Prevention and Control of Solid Waste Pollution

9.1 General Requirements

9.1.1 The stacking and storage of solid wastes shall comply with the provisions specified in national laws and regulations.

9.1.2 Pre-treatment measures for recycling and harmlessness shall be taken for solid wastes generated from railway operations.

9.2 Design for Prevention and Control of Solid Waste Pollution During Construction Period

9.2.1 Measures for domestic waste disposal during railway construction shall comply with the relevant provisions specified in national or local standards.

9.2.2 Construction wastes generated during the construction of railway projects in the urban planning area shall not be mixed with domestic wastes and shall be disposed in accordance with the provisions of the municipal competent authorities.

9.3 Design for Prevention and Control of Solid Waste Pollution During Operation Period

9.3.1 Coal slags generated from the coal-fired boiler room shall be utilized comprehensively and shall comply with relevant provisions specified in *Code for Design of Boiler Plant* (GB 50041).

9.3.2 Metal chips and other solid wastes generated from repair and processing of equipment in railway station/depot (post) shall be stacked and stored together as per categories, and should be recycled for comprehensive utilization.

9.3.3 Railway passenger stations, EMU depots (running sheds) and technical servicing depots for passenger trains shall be equipped with waste collection and transfer facilities as per the amount of wastes generated from passenger trains and the requirements of transportation organization. Other railway production units shall be equipped with waste collection and transfer facilities as per the amount of domestic wastes.

9.3.4 Hazardous solid wastes generated during the operation of the railway station/depot (post) shall be collected, stored and disposed in accordance with relevant national regulations.

10 Prevention and Control of Electromagnetic Pollution

10.0.1 The power-frequency electric field and the magnetic flux density outside the enclosure of the traction substation of 110 kV and above shall comply with the provisions specified in relevant technical standards such as *Controlling Limits for Electromagnetic Environment* (GB 8702).

10.0.2 The controlling limits for electromagnetic environment of base stations for the digital mobile communication system of railway shall comply with relevant provisions specified in *Controlling Limits for Electromagnetic Environment* (GB 8702).

10.0.3 Measures shall be taken when the received signal-to-noise ratio of TV for residential buildings on both sides of railway line is less than 35 dB due to the radio interference of electrified railway.

Words Used for Different Degrees of Strictness

In order to mark the differences in executing the requirements in this Code, words used for different degrees of strictness are explained as follows:

(1) Words denoting a very strict or mandatory requirement:

"Must" is used for affirmation; "must not" is used for negation.

(2) Words denoting a strict requirement under normal conditions:

"Shall" is used for affirmation; "shall not" is used for negation.

(3) Words denoting a permission of a slight choice or an indication of the most suitable choice when conditions permit:

"Should" is used for affirmation; "should not" is used for negation.

(4) "May" is used to express the option available, sometimes with the conditional permit.